AF588493

BIG CATS

LIONS

by Elizabeth Andrews

Cody Koala

An Imprint of Pop!

popbooksonline.com

Hello! My name is Cody Koala

This book is filled with videos, puzzles, games, and more! Scan the QR codes* while you read, or visit the website below to make this book pop.

*Scanning QR codes requires a web-enabled smart device with a QR code reader app and a camera.

abdobooks.com

Published by Pop!, a division of ABDO, PO Box 398166, Minneapolis, Minnesota 55439.

Printed in the United States of America, North Mankato, Minnesota.

102024
012025

THIS BOOK CONTAINS RECYCLED MATERIALS

Cover Photo: Shutterstock Images
Interior Photos: Getty Images, Shutterstock Images
Editor: Grace Hansen
Series Designer: Neil Klinepier, Candice Keimig

Library of Congress Control Number: 2024938601

Publisher's Cataloging-in-Publication Data

Names: Andrews, Elizabeth, author.
Title: Lions / by Elizabeth Andrews
Description: Minneapolis, Minnesota : Pop!, 2025 | Series: Big cats | Includes online resources and index
Identifiers: ISBN 9781098246921 (lib. bdg.) | ISBN 9781098247485 (ebook)
Subjects: LCSH: Big cats--Juvenile literature. | Wildcat--Juvenile literature. | Lion--Juvenile literature. | Lion--Behavior--Juvenile literature.
Classification: DDC 599.757--dc23

Table of Contents

Chapter 1

What Is a Lion?

A lion is a powerful big cat. Lions are called the "Kings of the Jungle." They have golden yellow fur. Adult male lions have manes that can be light tan to dark brown.

Lions' manes protect their necks during fights.
Watch a video here!

Lions have excellent eyesight. They can even see at night. Lions also have good hearing. They can hear their **prey** from more than a mile (1.6km) away!

Most lions in the world live in the grassy plains and **savannas** of Africa. There are a small number of lions living in India's Gir Forest.

Where Lions Live

Europe

Asia

Africa

South America

Atlantic Ocean

Indian Ocean

N
W
E
S

Lion Range

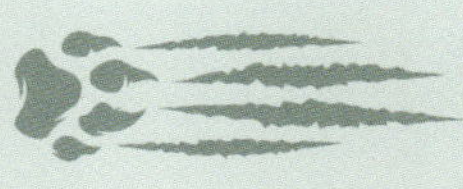

Chapter 2

The Pride

Lions live in groups called prides. Prides usually have 10 to 40 members. Each pride has one to three males, many females, and their cubs.

Learn more here!
Lions are the only big cats that live in groups.

Each pride has its own **territory**. Male lions protect the pride. They work to keep other lions and danger away. They roar and leave scent marks to warn others.

Chapter 3

Lion Habits

Prides live, play, hunt, sleep, and eat together. Lions are **carnivores**. They eat animals, such as zebras, wildebeest, and antelope.

Explore links here!

Lions also **scavenge** for food.

Female lions do the hunting. They often hunt in groups and work together to take down their **prey**.

Lions keep track of each other by watching for the dark tips of their long tails in the grass.

Chapter 4

Lion Cubs

Mother lions have one to six cubs at a time. Cubs are born with spots. The spots help them **camouflage** when their mother leaves them in tall grasses.

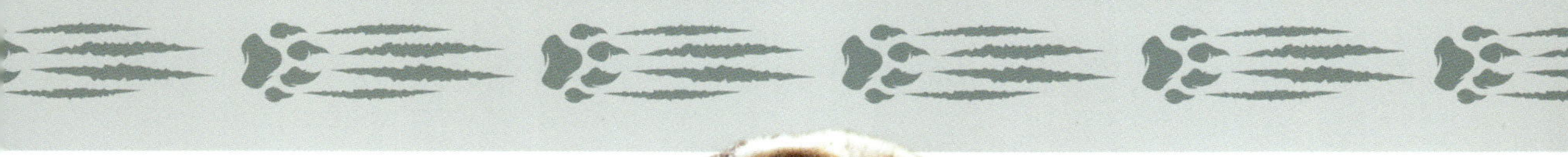

Complete an activity here!

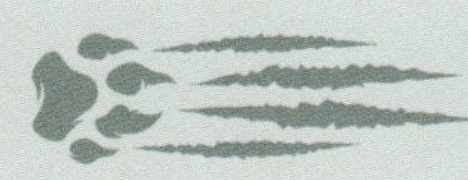
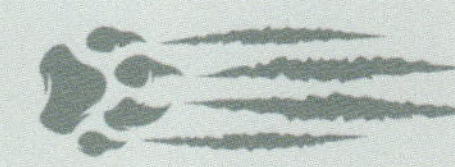

Female lion cubs stay with their pride forever. Male cubs stay for about two years until

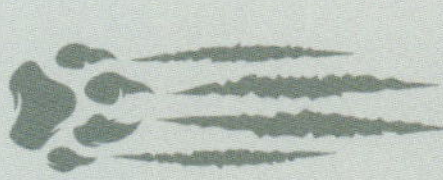
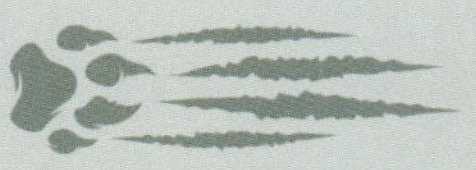

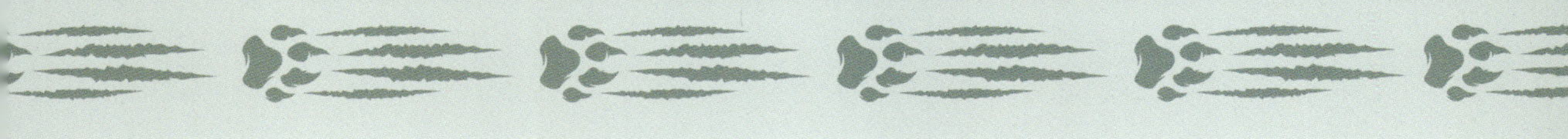

Lions live 10 to 15 years in the wild.

they are chased out of their pride. They travel to find a new pride to take over.

Making Connections

Text-to-Self

What big cat are you most interested in? Please explain your answer.

Text-to-Text

Have you read about any other kinds of big cats? If so, how were those big cats similar to or different from lions?

Text-to-World

Lion cubs have markings to help them hide. Can you think of any other cubs or baby animals that have markings that help them hide?

Glossary

camouflage – to hide by coloring or covering to look like the surroundings.

carnivore – an animal that feeds on other animals.

prey – an animal that is hunted by other animals for food.

savanna – flat grasslands with scattered trees and shrubs.

scavenge – to feed on already dead animals.

territory – a particular area of land that belongs to an animal.

Index

Online Resources

popbooksonline.com

Thanks for reading this Cody Koala book!

This book is filled with videos, puzzles, games, and more! Scan the QR codes* while you read, or visit the website below to make this book pop.

popbooksonline.com/lion

*Scanning QR codes requires a web-enabled smart device with a QR code reader app and a camera.